Student Solutions Manual for

Concepts in Probability and Stochastic Modeling

James J. Higgins
Kansas State University

Sallie Keller-McNulty
Kansas State University

Mary E. Muckenthaler
Kansas State University

Duxbury Press
An Imprint of Wadsworth Publishing Company
I(T)P™ An International Thomson Publishing Company

Belmont • Albany • Bonn • Boston • Cincinnati • Detroit • London • Madrid • Melbourne
Mexico City • New York • Paris • San Francisco • Singapore • Tokyo • Toronto • Washington

Printed in the United States of America
1 2 3 4 5 6 7 8 9 10—01 00 99 98 97 96 95

For more information, contact Wadsworth Publishing Company:

Wadsworth Publishing Company
10 Davis Drive
Belmont, California 94002, USA

International Thomson Publishing Europe
Berkshire House 168-173
High Holborn
London, WC1V 7AA, England

Thomas Nelson Australia
102 Dodds Street
South Melbourne 3205
Victoria, Australia

Nelson Canada
1120 Birchmount Road
Scarborough, Ontario
Canada M1K 5G4

International Thomson Editores
Campos Eliseos 385, Piso 7
Col. Polanco
11560 México D.F. México

International Thomson Publishing GmbH
Königswinterer Strasse 418
53227 Bonn, Germany

International Thomson Publishing Asia
221 Henderson Road
#05-10 Henderson Building
Singapore 0315

International Thomson Publishing Japan
Hirakawacho Kyowa Building, 3F
2-2-1 Hirakawacho
Chiyoda-ku, Tokyo 102, Japan

ISBN 0-534-23137-3

CONTENTS

CHAPTER 1
BASIC PROBABILITY

SECTION 1.1

1.1-1   **a** $S = \{(r,b), (r,y), (b,r), (b,y), (y,r), (y,b)\}$
       b   4

1.1-3   **a** $S = \{(1,2,3), (1,3,2), (2,1,3), (2,3,1), (3,1,2), (3,2,1)\}$
       b   2

1.1-5   **a** $S = \{0,1,2\}$
       b $S = \{(d,d), (d,g), (g,d), (g,g)\}$
       c $S = \{(1,2), (1,3), (1,4), (2,1), (2,3), (2,4), (3,1), (3,2), (3,4), (4,1), (4,2),$
          $(4,3)\}$
       d   For part a, $S = \{0\}$; For part b, $S = \{(g,g)\}$; For part c, $S = \{(3,4), (4,3)\}$
       e   For part a, $S = \{2\}$; For part b, $S = \{(0,0)\}$; For part c, $S = \{(1,2), (2,1)\}$

SECTION 1.2

1.2-1 $A^c = .7$; $B^c = .2$; $(A \cap B)^c = .8$

1.2-3   **a** $S = \{(1,2), (1,3), (1,4), (1,5), (1,6), (1,7), (1,8),$

          $(8,1), (8,2), (8,3), (8,4), (8,5), (8,6), (8,7)\}$
       b $6/56 = 3/28$
       c $20/56 = 5/14$
       d $30/56 = 15/28$
       e $21/56$
       f   No.
       g   No: P(2 defective chips) = 3/28; P(2 good chips) = 10/28
          P(1 good and 1 defective) = 15/28

1.2-7   **a** $S = \{(s,s), (s,f), (f,f), (f,s)\}$, where s = success and f = failure

b For the first trial, let 1, 2, or 3 = s and 4, 5 or 6 = f.
For the second trial, let 1,2,3 or 4 = s and 5 or 6 = f.

S (faces of the die)	S
(1,1) (1,2) (1,3) (1,4) (2,1) (2,2) (2,3) (2,4) (3,1) (3,2) (3;3) (3,4)	(s,s)
(1,5) (1,6) (2,5) (2,6) (3,5) (3,6)	(s,f)
(4,5) (4,6) (5,5) (5,6) (6,5) (6,6)	(f,f)
(4,1) (4,2) (4,3) (4,4) (5,1) (5,2)(5,3) (5,4) (6,1) (6,2) (6,3) (6,4)	(f,s)

Thus, $P(s,s) = 1/3$, $P(s,f) = 1/6$, $P(f,f) = 1/6$, and $P(f,s) = 1/3$.

c P(number of failed tests is exactly one) = $P((s,f) \cup (f,s)) = 18/36$

1.2-9 **a** Call the celebrities A, B, and C. To correctly match all of the photos, the contestant must arrange the photographs in the order ABC. Other combinations are ACB, BCA, BAC, CBA, and CAB. Note that there are 6 arrangements, each with probability 1/6.
P(all pictures match) = P(ABC) = 1/6.

b P(none of the pictures match) = P(BCA)+P(CAB) = 4/6 = 2/3.

SECTION 1.3

1.3-1 **a** For example, start at column 7, line 1, in Appendix 1 and designate the following:

Random number	Die roll
0 - .16666	1
.16667-.33333	2
.33334-.50000	3
.50001-.66666	4
.66667-.83333	5
.83334-.99999	6

Generated rolls = (4,2), (1,3), (4,2), (5,6), (4,3), (4,2), (3,5), (3,4), (4,2), (5,5), (6,5), (6,3),
(4,4), (6,2), (3,3), (5,5), (2,2), (2,2), (1,2), (1,4)
Proportion of tosses resulting in a 7 = 2/20

b P(observing a 7) = P((1,6), (6,1), (2,5), (5,2), (3,4), (4,3)) = 6/36 = 1/6

1.3-5 Description: This simulation was accomplished by generating a unit random number on [0,1] and comparing it to .5. If the random number < .5, then tail was tossed; otherwise, a head was tossed. S(·), the number of heads minus the number of tails for n tosses was recorded, and in addition, a running total of heads and tails and the maximum and minimum S(·) was recorded.

Variables:

h	number of heads
t	number of tails
$S(\cdot)$	an array containing the number of heads minus the number of tails for n tosses
c	number of times $S(\cdot)$ changes sign
max	maximum $S(\cdot)$
min	minimum $S(\cdot)$

Initialize:

$h = t = c = max = min = S(-1) = S(0) = 0$

Algorithm:

Repeat 5000 times
1) Generate u, a uniform random number on $[0,1]$
2) If $u < .5$, set $S(n) = S(n-1) - 1$ and $t = t + 1$
   else set $S(n) = S(n-1) + 1$ and $h = h + 1$
3) If $S(n) > max$, set $max = S(n)$
4) If $S(n) < min$, set $min = S(n)$
5) If $S(n-2) + S(n-1) + S(n) = 0$, then increment $c = c + 1$
6) Let $n = n + 1$

Sample Output:

For this simulation, there were 5000 trials.

Trial	1	2	3
frequency of heads	.494	.525	.492
frequency of tails	.506	.475	.508
# of sign changes	19	14	17
maximum accumulated money	26	52	9
minimum accumulated money	-7	-4	-38

SECTION 1.4

1.4-1 $(4)(3)(2)(6) = 144$

1.4-3   Number of words with at least 15 ones $= \begin{pmatrix} 16 \\ 15 \end{pmatrix} + 1 = 17$

Number of words with exactly 12 ones and 4 zeros $= \begin{pmatrix} 16 \\ 12 \end{pmatrix} \begin{pmatrix} 4 \\ 4 \end{pmatrix} = 1820$

1.4-7   **a** $(2)(9)(8)(7) = 1008$
             b $(2)(10)(10)(10) = 2000$
             c $(9)(10)(10)(10) = 9000$

3

1.4-13 $\dfrac{\dbinom{10}{7}}{\dbinom{15}{7}} = .0186$

1.4-15 Total number of ordered arrangements = n!
For each ordered arrangement, there are $r_1!r_2!...r_k!$ unordered arrangements.
The number of unordered arrangements $= \dfrac{n!}{r_1!\,r_2!\,...\,r_k!}$

Number of arrangements in Mississippi $= \dfrac{11!}{1!4!4!2!} = 34,650$

SECTION 1.5

1.5-1 Let A = event of a 3 and a 4 $\Rightarrow$ P(A) = 1/18
Let B = event total is a 7 $\Rightarrow$ P(B) = 1/6
Note that $A \subseteq B$ so that $P(A \cap B) = P(A)$.

$$P(A \mid B) = \dfrac{P(A \cap B)}{P(B)} = \dfrac{P(A)}{P(B)} = \dfrac{1/18}{1/6} = 1/3$$

1.5-9 Let B_1 = Box 1 selected
Let B_2 = Box 2 selected
Let C_B = Broken cookie selected
Let C_G = Good cookie selected
a $P(C_B) = P(B_1)P(C_B|B_1) + P(B_2)P(C_B|B_2) = (1/2)(6/10) + (1/2)(7/12) = .59$
b P(2 broken cookies selected|B_1)P(B_1) + P(2 broken cookies selected|B_2)P(B_2)

$$= \dfrac{\dbinom{6}{2}}{\dbinom{10}{2}}\,(1/2) + \dfrac{\dbinom{7}{2}}{\dbinom{12}{2}}\,(1/2) = (15/45)(1/2) + (21/66)(1/2) = .33$$

1.5-13 $P(E_J|A) = \dfrac{P(E_J \cap A)}{P(A)} = \dfrac{P(A|E_J)P(E_J)}{\sum P(A \cap E_J)} = \dfrac{P(A|E_J)}{\sum P(E_J)P(A|E_J)}$

SECTION 1.6

1.6-1 **a** Not independent--weather often occurs in patterns (e.g. a storm system).
b Independent--the failure of one should have no effect on the failure of the other.
c Not independent--salaries are often increased to reflect the increased cost of living.

d Not independent--Tossing 2 heads affects the occurrence of 2 tails.

1.6-3 **a** $P(A \cap B) = P(A|B)P(B) = (.5)(.25) = .125$
$P(A \cup B) = P(A) + P(B) - P(A \cap B)$
$\Rightarrow .75 = P(A) + .25 - .125$
$\Rightarrow P(A) = .625$
$P(B^C|A^C) = \dfrac{P(B^C \cap A^C)}{P(A^C)} = \dfrac{P(A \cup B)^C}{P(A^C)} = \dfrac{1 - .75}{1 - .625} = 2/3$

b A and B are not independent events since
$P(A \cap B) = .125$ and $P(A)P(B) = .15625$

1.6-9 P(system works) = P(at least one component works)
$= 1 - P(\text{all components fail}) = 1 - (1-p_1)(1-p_2)(1-p_3)$

1.6-11 P(bad item is detected at each inspection) = .75
P(bad item passes 3 inspections)
= P(bad item passes 1st)P(bad item passes 2nd)P(bad item passes third) =
$(.25)^3 = .015625$

CHAPTER 2
DISCRETE RANDOM VARIABLES

SECTION 2.1

2.1-1   **a**

x	2	3	4	5	6	7	8	9	10	11	12
p(x)	1/36	2/36	3/36	4/36	5/36	6/36	5/36	4/36	3/36	2/36	1/26

b

x	1	2	3	4	5	6
p(x)	1/36	3/36	5/36	7/36	9/36	11/36

2.1-3

y	0	20	40	60
p(y)	4/35	18/35	12/35	1/35

2.1-5   **a**

x	0	1	2	3	4
p(x)	0.2	0.3	0.3	0.1	0.1

b   (3 or more servers busy) = .20

SECTION 2.2

2.2-1   **a** $P(X > 2, Y > 1) = .1$
$P(Y \le 1) = .7$

b

y	0	1	2
p(y)	.3	.4	.3

c $P(Y = 0 | X = 0) = (.05)/(.15) = 1/3$

d   If x and y were independent, then $P_{XY}(x,y) = P_X(x)P_Y(y)$
However, $P_{XY}(0,0) = .05$ but $P_X(0)P_Y(0) = (.15)(.30) = .045 \ne .05$

2.2-3 **a**

	2	3	4	5	6	7	8	9	10	11	12
0	1/36	0	1/36	0	1/36	0	1/36	0	1/36	0	1/36
1	0	2/36	0	2/36	0	2/36	0	2/36	0	2/36	0
2	0	0	2/36	0	2/36	0	2/36	0	2/36	0	0
3	0	0	0	2/36	0	2/36	0	2/36	0	0	0
4	0	0	0	0	2/36	0	2/36	0	0	0	0
5	0	0	0	0	0	2/36	0	0	0	0	0

b $P(Y = 1 \mid X = 7) = \dfrac{(2/36)}{(2/36) + (2/36) + (2/36)} = 1/3$

SECTION 2.3

2.3-1 $E(X) = (1/6)(1 + 2 + 3 + 4 + 5 + 6) = (21/6)$

2.3-3 Since X = .85Y - 10, $E(X) = .85E(Y) - 10 = (.85)(100) - 10 = \$75$

SECTION 2.4

2.4-1 **a** $E(X) = (1/6)(1 + 2 + 3 + 4 + 5 + 6) = 21/6$
$VAR(X) = (1/6)[(1 - (21/6)^2) + (2 - (21/6)^2) + (3 - (21/6)^2) + (4 - (21/6)^2)$
$\quad + (5 - (21/6)^2) + (6 - (21/6)^2)] = 2.92$
$STD(X) = 1.71$

b A two standard deviation interval about the mean is $(21/6) \pm 2(1.71)$, or [.08,6.92]. All of the numbers 1-6 fall into this range. Thus 100% of the probability is contained in this interval.

2.4-3 **a** $E(X) = 0(.25) + 1(.25) + 2(.20) + 3(.10) + 4(.10) + 5(.10) = 1.85$
$VAR(X) = (.25)(0 - 1.85)^2 + (.25)(1 - 1.85)^2 + (.20)(2 - 1.85)^2$
$\quad + (.10)(3 - 1.85)^2 + (.10)(4 - 1.85)^2 + (.10)(5 - 1.85)^2 = 2.63$
$STD(X) = 1.62$

b Let Y = weekly cost
$Y = 350X + 100$
$E(Y) = 350E(X) + 100 = 747.5$
$VAR(Y) = VAR(350X + 100) = 350^2 VAR(X) = 321,868.8$
$STD(Y) = 567.3$

SECTION 2.5

2.5-1   Let r be a two-digit random number from the table.
If $r < 50$, set $Y = 0$; if $50 \leq r < 80$, set $Y = 70$; if $80 \leq r < 95$, set $Y = 140$; otherwise set $Y = 210$.

2.5-3   Description: To simulate weekly lost revenue for an airline, 5 random numbers on $[0,1]$ are generated and each corresponds to a value of Y, the daily lost revenue, according to the probability distribution of Y.
Variables:

S(·)	an array from 0 to 15 to record the number of times 0 - 15 vacant seats occur
u	a unit random number
Y	the amount of lost revenue

Algorithm:
Repeat 1000 times
   Repeat the next 2 steps 5 times
      1) Generate u, a unit random number
      2) If $u < .50$ then set $Y = Y + 0$ and vacseats = 0
           else if $u < .80$ then set $Y = Y + 70$
           else if $u < .95$ then set $Y = Y + 140$
           else set $Y = Y + 210$
   Increment S[Y/70]
    Set $Y = 0$
Sample Output:

y	0	70	140	210	280	350	420	490	560
p(y)	.029	.092	.163	.215	.219	.134	.073	.044	.021

y	630	700	770	840	910	980	1,050
p(y)	.009	0	.001	0	0	0	0

SECTION 2.6

2.6-1 $\overline{X} = 6.9$
$S^2 = (1/10)[(11 - 6.9)^2 + (9 - 6.9)^2 + (3 - 6.9)^2 + (7 - 6.9)^2 + (8 - 6.9)^2 + (1 - 6.9)^2$
    $+ (5 - 6.9)^2 + (12 - 6.9)^2 + (3 - 6.9)^2 + (10 - 6.9)^2] = 12.69$
$S = 3.56$

2.6-3 $\overline{X} = 3.6$
$S^2 = (1/15)[(5 - 3.6)^2 + (1 - 3.6)^2 + (2 - 3.6)^2 + (1 - 3.6)^2 + (2 - 3.6)^2$
    $+ (3 - 3.6)^2 + (6 - 3.6)^2 + (0 - 3.6)^2 + (4 - 3.6)^2 + (3 - 3.6)^2 + (7 - 3.6)^2$
    $+ (2 - 3.6)^2 + (1 - 3.6)^2 + (8 - 3.6)^2 + (9 - 3.6)^2] = 7.307$
$S = 2.703$

2.6-5 $S^2 = (1/10)(11^2 + 9^2 + 3^2 + 7^2 + 8^2 + 1^2 + 5^2 + 12^2 + 3^2 + 10^2 - 10(6.9)^2)$
$= 12.69$

2.6-7 **a** $E(X) = 3.3$
$VAR(X) = (1 - 3.3)^2(.1) + (2 - 3.3)^2(.2) + (3 - 3.3)^2(.2) + (4 - 3.3)^2(.3)$
$+ (5 - 3.3)^2(.2) = 1.61$
$STD(X) = 1.269$

b Description: This program generates a unit random number and uses the cumulative distribution function for x to assign values. The sample mean and standard deviation are computed for 10 and 100 variables.
Variables:

u	a unit random number
X	the simulated random variable
S(·)	an array recording the value of X
i	a loop counter

Algorithm:
Repeat 15 times
    Repeat the next 4 steps 10 (100) times
        1) Generate u, a unit random number
        2) If u < .1, then set X = 1
           else if u < .3, then set X = 2
           else if u < .5, then set X = 3
           else if u < .8, then set X = 4
           else set X = 5
        3) Let S[i] = X
        4) Increment i
        5) Compute the sample mean and standard deviation
Sample Output:
For 10 trials,
Sample Mean: 3.200
Sample Standard Deviation: 1.249
For 100 trials,
Sample Mean: 3.310
Sample Standard Deviation: 1.286

SECTION 2.7

2.7-1 **a** $E(X) = (2)(.37) + (3)(.33) + (4)(.30) = 2.93$
$E(Y) = (-1)(.35) + (0)(.52) + (1)(.13) = -.22$
$E(X - Y) = (X) = E(Y) = 3.15$
$E(2X - 3Y) = 2E(X) - 3E(Y) = 6.52$

b $E(XY) = (-2)(.10) + (-3)(.20) + (-4)(.05) + (3)(.03) + (4)(.10) = -.51$
$E(X)E(Y) = (2.93)(-.22 = -.6446$ which does not equal $E(XY)$.
Thus, X and Y are not independent.

2.7-5 $E(W) = E(X) + E(Y) = 2E(X) = 2[1/6 + 2/6 + 3/6 + 4/6 + 5/6 + 6/6]$
$= (2)(21/6) = 7$
$VAR(X + Y) = VAR(X) + VAR(Y) = 2VAR(X) = (2/6)[(1 - 3.5)^2 + (2 - 3.5)^2$
$+ (3.- 3.5)^2 + (4 - 3.5)^2 + (5 - 3.5)^2 + (6 - 3.5)^2] = 5.8333$
$STD(X + Y) = 2.415$

2.7-7 $E(W) = 262.22$ (values obtained from table)
$VAR(W) = (0 - 262.22)^2(.021) + (70 - 262.22)^2(.086) + (140 - 262.22)^2(.170)$
$+ (21 - 262.22)^2(.203) + (280 - 262.22)^2(.195) + (350 - 262.22)^2(.158)$
$+ (420 - 262.22)^2(.084) + (490 - 262.22)^2(.049) + (560 - 262.22)^2(.016)$
$+ (63 - 262.22)^2(.015) + (700 - 262.22)^2(.002) + (770 - 262.22)^2(.005)$
$= 17,715.87$
$STD(W) = 133.10$
$E(W) = E(\sum Y_i) = 5E(Y) = 5[(0)(.5) + (70)(.3) + (140)(.15) + (210)(.05)]$
$= (5)(52.5) = 262.5$
$VAR(W) = VAR(\sum Y_i) = 5VAR(Y)$
$= 5[(0 - 52.5)^2(.5) + (70 - 52.5)^2(.3) + (140 - 52.5)^2(.15 + (210 - 52.5)^2(.05)]$
$= 19,293.75$
$STD(W) = 138.9$

SECTION 2.8

2.8-1 **a** $E(X) = (0)(.13) + (1)(.46) + (2)(.28) + (3)(.13) = 1.41$
$VAR(X) = (0 - 1.41)^2(.13) + (1 - 1.41)^2(.46) + (2 - 1.41)^2(.28)$
$+ (3 - 1.41)^2(.13) = .7619$
$STD(X) = .8729$
$E(Y) = (0)(.3) + (1)(.4) + (2)(.2) + (3)(.1) = 1.1$
$VAR(Y) = (0 - 1.1)^2(.3) + (1 - 1.1)^2(.4) + (2 - 1.1)^2(.2) + (3 - 1.1)^2(.1) = .89$
$STD(Y) = .89$

b $E(X) = 1.41$
$E(Y) = 1.1$
$E(XY) = (0)(.37) + (1)(.20) + (2)(.20) + (3)(.07) + (4)(.06) + (6)(.08)$
$+ (9)(.02) = 1.71$
$COV(X,Y) = 1.71 - (1.41)(1.1) = .159$
$CORR(X,Y) = \dfrac{.159}{(.8729)(.9434)} = .1931$

c $Cost = (3.00)X + (3.25)Y$
$E(Cost) = (3.00)E(X) + (3.25)E(Y) = (3.00)(1.41) + (3.25)(1.1) = 7.805$
$VAR(Cost) = (3.00)^2VAR(X) + (3.25)^2VAR(Y) = (3.00)^2(.7619)$
$+ (3.25)^2(.89) = 16.2577$
$STD(Cost) = 4.032$

2.8-5 Length, height, and width of a piece of wood; number of leaves, trunk diameter, and number of lemons on a tree.

SECTION 2.9

2.9-1 $P(Y = -1 \mid X = 2) = .10/.37 = .27$
$P(Y = 0 \mid X = 2) = .27/.37 = .73$
$P(Y = 1 \mid X = 2) = 0/.37 =$
$E(Y \mid X = 2) = (-1)(.27) + (0)(.73) + (1)(0) = -.27$
$P(X = 2 \mid Y = 0) = .27/.52 = .52$
$P(X = 3 \mid Y = 0) = .10/.52 = .19$
$P(X = 4 \mid Y = 0) = .15/.52 = .29$
$E(X \mid Y = 0) = 2.77$

CHAPTER 3
SPECIAL DISCRETE RANDOM VARIABLES

SECTION 3.1

3.1-1 $P(Y = 0) = \begin{pmatrix} 8 \\ 0 \end{pmatrix} (.25)^0 (.75)^8 = .1001$

$P(Y = 1) = \begin{pmatrix} 8 \\ 1 \end{pmatrix} (.25)^1 (.75)^7 = .2670$

$P(2 \le Y \le 4) = \begin{pmatrix} 8 \\ 2 \end{pmatrix} .25^2 .75^6 + \begin{pmatrix} 8 \\ 3 \end{pmatrix} .25^3 .75^5 + \begin{pmatrix} 8 \\ 4 \end{pmatrix} .25^4 .75^4 = .6056$

$P(Y > 2) = 1 - [P(Y = 0) + P(Y = 1) + P(Y = 2)] = 1 - [.1001 + .2670 + .3114]$
$= .3215$

3.1-3 $P(Y \ge 4) = \begin{pmatrix} 5 \\ 4 \end{pmatrix} .25^4 .75^1 + \begin{pmatrix} 5 \\ 5 \end{pmatrix} .25^5 .75^0 = .01562$

3.1-5   **a**   Let Y = the event a 1 occurs
Y is Bernoulli with P(success)= 1/6
E(Y) = np = 600/6 = 100
VAR(Y) = np(1-p) = 600(1/6)(5/6) = 83.33
STD(Y) = 9.1287
b $100 \pm 2(9.1287)$

3.1-7 $P(\text{customer 11 waits}) = \begin{pmatrix} 10 \\ 10 \end{pmatrix} .6^{10} .4^0 = .0060$

$P(\text{customer 12 waits}) = \begin{pmatrix} 11 \\ 10 \end{pmatrix} .6^{10} .4^1 = .0266$

$P(\text{customer 13 waits}) = \begin{pmatrix} 12 \\ 10 \end{pmatrix} .6^{10} .4^2 = .0638$

$P(\text{customer 14 waits}) = \begin{pmatrix} 13 \\ 10 \end{pmatrix} .6^{10} .4^3 = .1106$

$P(\text{customer 15 waits}) = \begin{pmatrix} 14 \\ 10 \end{pmatrix} .6^{10} .4^4 = .1549$

$P(\text{customer waits}) = .0060 + .0266 + .0638 + .1106 + .1549 = .3619$

SECTION 3.2

3.2-1 **a** $P(x) = (2/3)^{x-1}(1/3)$

   b $P(X = 5) = (2/3)^{5-1}(1/3) = .0658$

   c $P(X \geq 4) = 1 - [P(X = 1) + P(X = 2) + P(X = 3)]$
   $= 1 - [1/3 + (2/3)(1/3) + (2/3)^2(1/3)] = .2963$

   d $E(X) = 1/(1/3) = 3$, where success is defined as when a miss occurs

3.2-3 **a** $P(X = 3) = (.5)^3 = .125$

   $$P(X = 4) = \begin{pmatrix} 3 \\ 2 \end{pmatrix}(.5)^3(.5) = .1875$$

   $$P(X = 5) = \begin{pmatrix} 4 \\ 2 \end{pmatrix}(.5)^3(.5)^2 = .1875$$

   b $VAR(X) = (3)(1 - .5)/(.5)^2 = 6$
   $STD(X) = 2.449$
   $E(X) = (3)/(.5) = 6$
   A likely range is $6 \pm 2(2.449)$, or $(1, 11)$ children

3.2-7 $E(Y) = 3/(.4) = .7.5$
   $VAR(Y) = 3(1 - .4)/(.4)^2 = 11.25$
   $STD(Y) = 3.354$

SECTION 3.3

3.3-1 **a** $0 \leq X \leq 9$

   b $E(X) = n(r/N) = (12)(9/20) = 5.4$
   $VAR(X) = n(r/N)(1 - r/N)((N - n)/(N - 1))$
   $= (12)(9/20)(1 - 9/20)((20 - 12)/(20 - 1)) = 1.25$

   c $0 \leq X \leq 5$
   $E(X) = (12)(5/20) = 3$
   $VAR(X) = (12)(5/20)(1 - 5/20)((20 - 12)/(20 - 1)) = .9474$

3.3-5 $$P(X \leq 1) = P(X = 0) + P(X = 1) = \frac{\begin{pmatrix} 10 \\ 0 \end{pmatrix}\begin{pmatrix} 10 \\ 8 \end{pmatrix}}{\begin{pmatrix} 20 \\ 8 \end{pmatrix}} + \frac{\begin{pmatrix} 10 \\ 1 \end{pmatrix}\begin{pmatrix} 10 \\ 7 \end{pmatrix}}{\begin{pmatrix} 20 \\ 8 \end{pmatrix}} = .0098$$

There is a very low probability that the auditor would obtain 0 or 1 delinquent accounts given this situation. This result is not consistent with random selection of accounts.

SECTION 3.4

3.4-1 $\dfrac{30!}{20!5!4!1!}(.15)^5(.07)^4(.03)^1(.75)^{20} = .0065$

3.4-3 Expected repair cost for one chip = $(0)(.65) + (10)(.22) + (30)(.13) = 6.1$
Expected repair cost for fifty chips = $(50)(6.1) = 305$
Variance of the cost = $(0)(50)(.65)(.35) + (100)(50)(.22)(.78)$
$\quad + (900)(50)(.13)(.87) + (2)(10)(30)(-50)(.22)(.13) = 5089.5$
Standard deviation = 71.34

SECTION 3.5

3.5-1 $P(X = 0) = \dfrac{e^{-2}2^0}{0!} = .1353$

$P(X = 1) = \dfrac{e^{-2}2^1}{1!} = .2707$

$P(2 \le X \le 4) = \dfrac{e^{-2}2^2}{2!} + \dfrac{e^{-2}2^3}{3!} + \dfrac{e^{-2}2^4}{4!} = .5413$

$P(X \ge 2) = 1 - P[(X = 0) + P(X = 1)] = 1 - (.1353 + .2707) = .5940$

3.5-3 $\mu \approx np = (20)(.02) = .4$
a $P(X = 0) = \dfrac{e^{-.4}(.1)^0}{0!} = .6703$
b $P(X \ge 2) = 1 - [P(X = 0) + P(X = 1)] = .0616$

3.5-5 $E(X) = \mu = 30$
$VAR(X) = \mu = 30$
A likely range is given by $30 \pm 2\sqrt{30}$

SECTION 3.6

3.6-1 **a** $E(X) = \dfrac{d}{dt} M_x(t)\big|_{t=0} = \dfrac{d}{dt}(1 - p + pe^t)\big|_{t=0} = pe^t\big|_{t=0} = p$

$E(X^2) = \dfrac{d^2}{dt^2} M_x(t)\big|_{t=0} = \dfrac{d^2}{dt^2}(1 - p + pe^t) = pe^t\big|_{t=0} = p$

$VAR(X) = p - p^2 = p(1 - p)$

b $E(X^k) = \dfrac{d^k}{dt^k}(1 - p + pe^t)\big|_{t=0}$

3.6-3 $M_x(t) = Ee^{zt} = Ee^{(x+y)t} = Ee^{xt}e^{yt} = Ee^{xt}Ee^{yt} = M_x(t)M_y(t)$

CHAPTER 4
MARKOV CHAINS

SECTION 4.1

4.1-5   Description: This program models the operation of the Little Green Computing
         Machine Company, which has 2 technicians and 10 computers, each with a
         probability of .2 of failing.
         Variables:
         DAY   The day of the week labeled 1,2,3,4,5
         BACKLOG(·)   An array consisting of the backlogs at the end of each
                                     day
         GOOD   The number of good computers at the beginning of
                                     each day
         FAIL   The number of computers that fail during the day
         IDLEDAYS   The number of days in the week the technicians are
                                     idle
         Initialize:
         IDLEDAYS = FAIL = BACKLOG(·) = 0
         GOOD = 10
         Algorithm:
         Repeat 1000 times
             Repeat the following steps for DAY = 1 to 5
                 1)   Determine the number of computers that FAIL during the day by
                      generating a unit
                      random number for each GOOD computer and determining the
                      number that are less
                      than .2.
                 2)   Determine the backlog at the end of the day. This is done as
                      follows: If BACKLOG(DAY - 1) = 0 or 1 and FAIL = 0 or 1, then
                      BACKLOG(DAY) = 0; elseBACKLOG(DAY)
                      = BACKLOG(DAY - 1) + FAIL - 2
                 3)   Update the number of GOOD computers as follows: GOOD = 10 -
                      BACKLOG(DAYS)
         Sample Output: (mean and standard deviation of number on backlog)

	Monday	Tuesday	Wednesday	Thursday	Friday
$\overline{X}$	.744	.705	.738	.744	.801
S	.781	.741	.784	.811	.842

Empirical probabilities for the number on backlog:

	0	1	2	3	4	5	6	7	8	9
Mon	.684	.195	.085	.032	.003	.001	.000	.000	.000	.000
Tues	.557	.227	.144	.048	.019	.005	.000	.000	.000	.000
Wed	.495	.253	.140	.072	.027	.013	.000	.000	.000	.000
Thurs	.487	.236	.168	.071	.028	.009	.001	.000	.000	.000
Fri	.457	.245	.166	.095	.025	.008	.004	.000	.000	.000

SECTION 4.2

4.2-1   The one-step transition matrix is given by

$$
P = \begin{array}{c} \\ 1 \\ 2 \\ 3 \\ 4 \\ 5 \\ 6 \end{array}
\begin{array}{c}
\begin{array}{cccccc} 1 & 2 & 3 & 4 & 5 & 6 \end{array} \\
\left[\begin{array}{cccccc}
0 & 1/2 & 0 & 0 & 0 & 1/2 \\
1/2 & 0 & 1/2 & 0 & 0 & 0 \\
0 & 1/2 & 0 & 1/2 & 0 & 0 \\
0 & 0 & 1/2 & 0 & 1/2 & 0 \\
0 & 0 & 0 & 1/2 & 0 & 1/2 \\
1/2 & 0 & 0 & 0 & 1/2 & 0
\end{array} \right]
\end{array}
$$

4.2-3   The one-step transition matrix is given by

$$
P = \begin{array}{c} \\ 0 \\ 1 \\ 2 \\ 3 \\ \vdots \end{array}
\begin{array}{c}
\begin{array}{ccccccc} 0 & 1 & 2 & 3 & 4 & \cdots \end{array} \\
\left[\begin{array}{ccccccc}
0 & 1 & 0 & 0 & 0 & \cdots \\
1/k & 0 & (k-1)/k & 0 & 0 & \cdots \\
0 & 2/k & 0 & (k-2)/k & 0 & \cdots \\
0 & 0 & 3/k & 0 & (k-3)/k & \cdots \\
\vdots & \vdots & \vdots & \vdots & \vdots & \ddots
\end{array} \right]
\end{array}
$$

4.2-7   The one-step transition matrix is given by

$$
P = \begin{array}{c} \\ 1 \\ 2 \\ 3 \\ 4 \\ 5 \\ 6 \end{array}
\begin{array}{c}
\begin{array}{cccccc} 1 & 2 & 3 & 4 & 5 & 6 \end{array} \\
\left[\begin{array}{cccccc}
P(A) & P(C) & P(B) & 0 & 0 & 0 \\
P(B) & P(A) & 0 & 0 & P(C) & 0 \\
P(A) & 0 & P(B) & P(C) & 0 & 0 \\
0 & 0 & P(A) & P(B) & 0 & P(C) \\
0 & P(A) & 0 & 0 & P(C) & P(B) \\
0 & 0 & 0 & P(B) & P(A) & P(C)
\end{array} \right]
\end{array}
$$

SECTION 4.3

4.3-1 **a** $P(1 \to 1 \to 1 \to 1 \to 0 \to 0) = (P(1 \to 1))^3 \, P(1 \to 0)P(0 \to 0) = .0032$

       b The two-step transition matrix is given by

$$P^2 = \begin{array}{cc} & \begin{array}{cc} 0 & 1 \end{array} \\ \begin{array}{c} 0 \\ 1 \end{array} & \left[\begin{array}{cc} .65 & .35 \\ .56 & .44 \end{array} \right] \end{array}$$

Thus, if the machine is down this hour, the probability that it is up two hours later is .35.

4.3-3   The one-step transition matrix is given by

$$P = \begin{array}{c} \\ 1 \\ 2 \\ 3 \\ 4 \end{array} \begin{array}{c} \begin{array}{cccc} 1 & 2 & 3 & 4 \end{array} \\ \left[\begin{array}{cccc} 0 & 1/3 & 1/3 & 1/3 \\ 1/2 & 0 & 0 & 1/2 \\ 1/2 & 0 & 0 & 1/2 \\ 1/3 & 1/3 & 1/3 & 0 \end{array} \right] \end{array}$$

Also, $\pi(0) = [1/4, 1/4, 1/4, 1/4]$

  a Since $\pi(0)P^1 = [1/3 \quad 1/6 \quad 1/6 \quad 1/3]$,

P(rat is in compartment 1 after one move) = 1/3
P(rat is in compartment 2 after one move) = 1/6
P(rat is in compartment 3 after one move) = 1/6
P(rat is in compartment 4 after one move) = 1/3

  b Since $\pi(0)P^2 = [.278 \quad .222 \quad .222 \quad .278]$,

P(rat is in compartment 1 after two moves) = .278
P(rat is in compartment 2 after two moves) = .222
P(rat is in compartment 3 after two moves) = .222
P(rat is in compartment 4 after two moves) = .278

4.3-7 **a** The three-step transition matrix is given by

$$P = \begin{array}{c} A \\ B \\ C \end{array} \left[\begin{array}{ccc} .486 & .269 & .245 \\ .378 & .295 & .327 \\ .296 & .310 & .394 \end{array} \right]$$

The probability that the student has an A at the end of the third course given that he or she has a B at the beginning of the first course is .378.

  b Since $\pi(0) = [.3 \ .5 \ .2]$, the probability distribution of grades at the end of the three-course sequence is given by $\pi(0)P^3 = [.3940 \ .2902 \ .3158]$.

SECTION 4.4

4.4-1   **a** $P(0 \to 0) = P(0 \text{ break down or } 1 \text{ breaks down})$

$$= \binom{2}{0}(.20)^0(.80)^2 + \binom{2}{1}(.20)^1(.80)^1 = .64 + .32 = .96$$

$$P(0 \to 1) = P(2 \text{ break down}) = \binom{2}{2}(.20)^2(.80)^0 = .04$$

$$P(1 \to 0) = P(0 \text{ break down}) = \binom{1}{0}(.20)^0(.80)^1 = .80$$

$$P(1 \to 1) = P(1 \text{ breaks down}) = \binom{1}{1}(.20)^1(.80)^0 = .20$$

Thus, the one-step transition matrix is given by

$$P = \begin{matrix} 0 \\ 1 \end{matrix} \begin{bmatrix} .96 & .04 \\ .80 & .20 \end{bmatrix}$$

b $P^2 = \begin{bmatrix} .9536 & .0464 \\ .9280 & .0720 \end{bmatrix}$ $P^3 = \begin{bmatrix} .9526 & .0474 \\ .9485 & .0515 \end{bmatrix}$

$P^4 = \begin{bmatrix} .9524 & .0476 \\ .9518 & .0482 \end{bmatrix}$ $P^5 = \begin{bmatrix} .9524 & .0476 \\ .9523 & .0477 \end{bmatrix}$

The number of computers on backlog are as follows:

	0	1
Monday	.9600	.0400
Tuesday	.9536	.0464
Wednesday	.9526	.0474
Thursday	.9524	.0476
Friday	.9524	.0476

4.4-3 $P^6 = \begin{bmatrix} .5025 & .3024 & .1516 & .0396 & .0040 \\ .4820 & .3075 & .1619 & .0440 & .0046 \\ .4415 & .3173 & .1823 & .0531 & .0056 \\ .3823 & .3303 & .2127 & .0670 & .0076 \\ .3071 & .3439 & .2526 & .0862 & .0103 \end{bmatrix}$

This matrix gives the probability of going from state i to state j after 6 working days.

SECTION 4.5

4.5-7   Description: This program is similar to exercise 4.5-6 but calculates the mean number of transitions it takes for the conversation to return to A. The mean, standard deviation and probability mass function are determined based on 1000 trials.

Variables:

SPEAKER	The state of the chain, or the person currently speaking
NT	A counter recording the number of transitions to return to A
S(·)	An array recording the number of transitions in each trial
n	A loop counter

Initialize:
SPEAKER = A, NT = 0

Algorithm:
Repeat 1000 times:
   Repeat the following steps until n > 0 and SPEAKER = A
      1)   Increment NT
      2)   Generate a unit random number u
      3)   Change SPEAKER based upon the one-step transition matrix and u
      4)   Increment n
   Set S(n) = NT
   Set n = 0
Calculate the sample mean, standard deviation, and probability mass function of S(·).

Sample Output: (For five trials of 1000 completed conversations)

Mean Number of Transitions	3.08	2.99	2.90	3.05	2.96
Sample Standard Deviation	1.34	1.38	1.24	1.51	1.41

Probability mass function:

Transitions	2	3	4	5	6	7	8	9	10	11
Probability	.528	.229	.123	.065	.030	.016	.005	.002	.000	.002

SECTION 4.6

4.6-1 $P^{10} = \begin{matrix} 1 \\ 2 \end{matrix} \begin{bmatrix} .6250 & .3750 \\ .6250 & .3750 \end{bmatrix}$,

where 1 corresponds to a sunny day, and 2 corresponds to a cloudy day. The steady-state vector is $\pi = (.6250, .3750)$. $\pi_1 = .6250$ is the fraction of days that will be sunny. π_2 is the fraction of days that will be cloudy.

4.6-3 $P^{100} = \begin{bmatrix} .4656 & .3800 & .1544 \\ .4656 & .3800 & .1544 \\ .4656 & .3800 & .1544 \end{bmatrix}$

The steady-state vector is given by $\pi = (.4656, .3800, .1544)$.

a In the long run, the taxicab will spend .4656 of the time at the airport, .3800 of the time at Hotel A, and .1544 of the time at Hotel B.

b If the taxicab begins at the airport, then the expected number of time periods that it will take for it to return to the airport is $1/\pi_1 = 1/.4656 = 2.14778$.

4.6-5 $\pi P = \pi \Rightarrow (\pi_0, \pi_1) \begin{bmatrix} 1-a & a \\ b & 1-b \end{bmatrix} = (\pi_0, \pi_1)$

$\Rightarrow (\pi_0, \pi_1) = (\pi_0 - \pi_0 a + \pi_1 b, \pi_0 a + \pi_1 - \pi_1 b)$

$\pi_0 - \pi_0 a + \pi_1 b = \pi_0$

$\pi_0 a + \pi_1 - \pi_1 b = \pi_1$

$\pi_0 + \pi_1 = 1 \Rightarrow \pi_1 = 1 - \pi_0$

From these equations, it follows that

$\pi_0 - \pi_1 - \pi_0 a + (1 - \pi_0)b = 0$

$\Rightarrow b - \pi_0 b - \pi_0 a = 0$

$\Rightarrow b = (a + b)\pi_0$

$\Rightarrow \pi_0 = b/(a + b)$

$\Rightarrow \pi_1 = 1 - b/(a + b) \Rightarrow \pi_1 = a/(a + b)$

SECTION 4.7

4.7-1 $P^3 = \begin{array}{c} 0 \\ 1 \\ 2 \\ 3 \end{array} \begin{bmatrix} 0 & .14 & .14 & .72 \\ 0 & 0 & .196 & .804 \\ 0 & .12 & 0 & .888 \\ 0 & 0 & 0 & 1 \end{bmatrix}$

a $P(T_0 \le 3) = P^{(3)}(0 \to 3) = .72$

b $P(T_0 = 2) = P(0 \to 1)P(1 \to 3) + P(0 \to 2)P(2 \to 3)$

$= (.5)(.3) + (.5)(.6) = .45$

4.7-5 **a** The one-step transition matrix is given by

$P = \begin{array}{c} H \\ AP \\ ANP \\ BP \\ BNP \\ CP \\ CNP \end{array} \begin{bmatrix} 0 & .6 & .4 & 0 & 0 & 0 & 0 \\ 0 & 1 & 0 & 0 & 0 & 0 & 0 \\ 0 & 0 & 0 & .6 & .4 & 0 & 0 \\ 0 & 0 & 0 & 1 & 0 & 0 & 0 \\ 0 & 0 & 0 & 0 & 0 & .3 & .7 \\ 0 & 0 & 0 & 0 & 0 & 1 & 0 \\ 0 & .6 & .4 & 0 & 0 & 0 & 0 \end{bmatrix}$

b The average number of transitions it takes for the commuter to find a parking place is found by making states AP, BP, and CP absorbing and solving the equation $U = (I - Q)^{-1} ONE$ where ONE is a column vector of ones, I is the identity matrix, and

$$Q = \begin{array}{c} H \\ ANP \\ BNP \\ CNP \end{array} \begin{bmatrix} 0 & .4 & 0 & 0 \\ 0 & 0 & .4 & 0 \\ 0 & 0 & 0 & .3 \\ 0 & .4 & 0 & 0 \end{bmatrix} . \text{ Since } (I - Q)^{-1} ONE = \begin{bmatrix} 1.608 \\ 1.52 \\ 1.30 \\ 1 \end{bmatrix} \text{ the}$$

average number of transitions the commuter must make is 1.608.

CHAPTER 5
CONTINUOUS RANDOM VARIABLES

SECTION 5.1

5.1-1 **a** $1 = \int_0^4 cx\,dx = \frac{cx^2}{2} \Big|_0^4 = 8c \Rightarrow c = 1/8$

b $P(1 \leq X \leq 2) = \int_1^2 (x/8)dx = x^2/16 \Big|_1^2 = 3/16$

5.1-3 **a** $\int_{1.5}^{\infty} \frac{3}{(x+1)^4}\,dx = \frac{3}{-3(x+1)^3} \Big|_{1.5}^{\infty} = .064$

b $1/2 = \int_0^m \frac{3}{(x+1)^4}\,dx = \frac{-1}{(x+1)^3} ;\Big|_0^m = \frac{-1}{(m+1)^3} - (-1) = 1 - \frac{1}{(m+1)^3}$

$\Rightarrow \frac{1}{(m+1)^3} = 1/2 \Rightarrow (m+1)^3 = 2 \Rightarrow m = 2^{1/3} - 1 \Rightarrow m = .2599$

5.1-5 **a** $1 - e^{-(.2)(10)} - (1 - e^{-(.2)(5)}) = e^{-(.2)(5)} - e^{-(.2)(10)} = .2325$
  b $1 - F(10) = e^{-.2(10)} = .1353$
  c $f(x) = .2e^{-.2x} \qquad x > 0$

SECTION 5.2

5.2-1 **a** $E(X) = \int_0^2 (x)(1 - x/2)dx = [(x^2/2) - (x^3/6)] \Big|_0^2 = 2 - 8/6 = 2/3$

$E(X^2) = \int_0^2 (x^2)(1 - x/2)dx = [(x^3/3) - (x^4/8)] \Big|_0^2 = (8/3) - (16/8) = 2/3$

$VAR(X) = (2/3) - (2/3)^2 = 4/9$
$STD(X) = 2/3$

b $2/3 \pm 2(2/3) = (-2/3, 2)$
$P(-2/3 < X < 1) = 1$

5.2-3 $E(X) = \int_{10}^{30} x\left(\frac{x-10}{200}\right)dx = (1/200)\left(\frac{x^3}{3} - \frac{10x^2}{2}\right)\Big|_{10}^{30} = 23.33$

$E(X^2) = \int_{10}^{30} x^2\left(\frac{x-10}{200}\right)dx = (1/200)\left(\frac{x^4}{4} - \frac{10x^3}{3}\right)\Big|_{10}^{30} = 566.66$

$$VAR(X) = 566.66 - (22.33)^2 = 22.22$$
$$STD(X) = 4.714$$

5.2-5 $f(x) = 1 - x/2$ $\qquad 0 \le x \le 2$
$\qquad\qquad\quad 0$ $\qquad\qquad\qquad$ otherwise

$Y = 2X - 3 \Rightarrow (Y + 3)/2 \Rightarrow \left|\dfrac{dx}{dy}\right| = 1/2$

$g(y) = [1 - \left(\dfrac{y+3}{2}\right)(1/2)](1/2) = (1 - y/4 - 3/4)(1/2)$

$= 1/8 - y/8 = 1/8(1 - y), \ -3 \le y \le 1$

$G(y) = \displaystyle\int_{-3}^{y} (1/8 - y/8)dy = (y/8 - y^2/16)|_{-3}^{y} = y/8 - y^2/16 + 3/8 + 9/16$

$= (1/16)[15 + 2y - y^2]$

SECTION 5.3

5.3-1 Description: This program simulates 1000 values from the distribution
$f(x) = 2x$. For each x, a unit random number is generated and set equal to $F(x)$.
Solving for x, we obtain $x = \sqrt{u}$.

Variables:

X	The simulated numbers
S(·)	An array of size 1000 containing the simulated values of X
n	A loop counter

Initialize:

$n = 1$

Algorithm:

Repeat 1000 times

    1)   Generate u, a unit random number

    2)   Set $X = \sqrt{u}$

    3)   Set $S(n) = X$

    4)   Set $n = n + 1$

Record the frequencies of X values for the 10 subintervals of length .1 and
construct a probability histogram.

Sample Output:

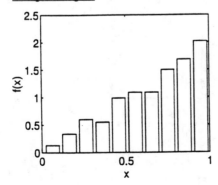

23

5.3-7   Description: This program simulates the appraised value of 1000 houses by using the mean value of 3 appraisals and the procedure described in the chapter.

Variables:

X, Y, W	The three appraised values
U	The average of the appraisals
S(·)	An array to hold the average appraisal value
n	A loop counter

Initialize:

n = 1

Algorithm:

Repeat 100 times

    1) Generate three random numbers, t, u, and v
    2) Set X = 20000t + 90,000
    3) Set Y = 25000u + 95,000
    4) Set W = 20000 $\sqrt{v}$ + 90,000
    5) Set U = (X + Y + W)/3
    6) S(n) = U
    7) Set n = n + 1

Construct a histogram.

Sample Output:

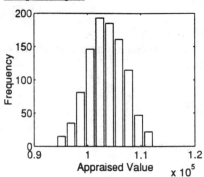

SECTION 5.4

5.4-1   a

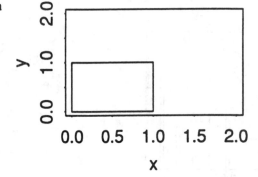

24

b $1 = \int_0^1\int_0^1 kdxdy = k \Rightarrow k = 1$

c $P(X < .5, Y > .5) = \int_0^{.5}\int_{.5}^1 1dxdy = 1/4$

d $f(x) = \int_0^1 1dy = 1, \; f(y) = \int_0^1 1dx = 1$

e yes, since $f(x)f(y) = f(x,y)$

5.4-3 **a** $P(X < .5, Y > .5) = \int_0^{1/2}\int_{1/2}^{1-x}(1/2)dydx = \int_0^{1/2}\left(\frac{1-x}{2}\right) - \left(\frac{1}{4}\right)dx$

$= \frac{x}{2} - \frac{x^2}{4} - \frac{x}{4} \Big|_0^{1/2} = 1/4 - 1/16 - 1/8 = .0625$

b $P(X > .75) = (1/2)(.25)(1 - .75)(1/2) = .01562$

c $P(X + Y < .25) = (1/2)(.25)(.25)(1/2) = .01562$

5.4-5 **a** $P(X < .5, Y > 1) = \int_0^{1/2}\int_1^2 (1/3)(x + y)dydx = \int_0^{1/2}(1/3)(xy + y^2/2)\Big|_1^2 \, dx$

$= \int_0^{1/2}(1/3)[2x + 2 - x - 1/2]dx = (1/3)\int_0^{1/2}(x + 3/2)dx = (1/3)(x2 + (3/2)x)\Big|_0^{1/2}$

$= (1/3)(1/8 + 3/4) = 7/24$

b $P(X < Y) = 1 - P(X \geq Y) = 1 - \int_0^1\int_0^x (1/3)(x + y)dydx$

$= 1 - \int_0^1 (1/3)(xy + y^2/2)\Big|_0^x \, dx = 1 - \int_0^1 (1/3)(x2 + x2/2)dx = 1 - \int_0^1 (x^2/2)dx$

$= 1 - (x^3/2)\Big|_0^1 = 5/6$

5.4-11 **a** $f(y_1) = \int_0^\infty \left(\frac{\exp(-\sqrt{y_1} - \sqrt{y_2})}{4\sqrt{y_1 y_2}}\right) dy_2 = \int_0^\infty \left(\frac{\exp(-\sqrt{y_1})}{2\sqrt{y_1}}\right)\left(\frac{\exp(-\sqrt{y_2})}{2\sqrt{y_2}}\right) dy_2$

$= \left(\frac{\exp(-\sqrt{y_1})}{2\sqrt{y_1}}\right)\exp(-\sqrt{y_2})\Big|_0^\infty = \left(\frac{\exp(-\sqrt{y_1})}{2\sqrt{y_1}}\right)$ for $y_1 > 0$

Similarly, $f(y_2) = \left(\frac{\exp(-\sqrt{y_2})}{2\sqrt{y_2}}\right)$

b Yes, since $f(y_1,y_2) = f(y_1)f(y_2)$

CHAPTER 6
SPECIAL CONTINUOUS RANDOM VARIABLES

SECTION 6.1

6.1-1   **a**   Since $\lambda = 2$, $E(X) = 1/2$
         b $STD(X) = 1/2$
         c $F(x) = 1 - e^{-2x}$, $0 < x < \infty$

6.1-3   **a** $E(X + Y) = E(X) + E(Y) = 2.5 + 3.5$
            $VAR(X + Y) = VAR(X) + VAR(Y) = (2.5)^2 + (3.5)^2 = 18.5$
         b $f(x,y) = \dfrac{1}{8.75}e^{-x/2.5}e^{-y/3.5}$, $0 < x < \infty$, $0 < y < \infty$

         c $P(X + Y > 15) = 1 - P(X + Y \le 15) = 1 - \dfrac{1}{8.75} \displaystyle\int_{y=0}^{15} \int_{x=0}^{15-y} e^{-x/2.5}e^{-y/3.5}dxdy$

$$= 1 - \int_{y=0}^{15} \frac{1}{3.5}(1 - e^{-(15-y)/2.5})(e^{-y/3.5})dy$$

$$= 1 - \int_{y=0}^{15} \frac{1}{3.5} e^{-y/3.5}\,dy + \int_{y=0}^{15} \frac{1}{3.5} e^{-6}\,e^{y/8.75}dy = e^{-y/3.5} \Big|_{0}^{15} + 2.5e^{-6}(e^{-x/8.75})\Big|_{0}^{15}$$

$$= 1 - (1 - e^{-15/3.5}) + 2.5e^{-6}(e^{15/8.75} - 1) = .042$$

6.1-5   Since $E(X) = 1/\lambda$ and $STD(X) = 1/\lambda$,
           $P(1/\lambda - 2/\lambda < X < 1/\lambda + 2/\lambda) = P(-1/\lambda < X < 3/\lambda) = P(0 < X < 3/\lambda)$
           $= F(3/\lambda) = 1 - e^{-3} = .9502$

SECTION 6.2

6.2-1   **a** $P(Z \le -.52) = 1 - .6985 = .3015$
         b $P(Z \le .52) = .6985$
         c $P(-.52 \le Z \le .52) = P(Z \le .52) - P(Z \le -.52) = .6985 - .30 = .3970$
         d $P(Z > 1.91) = .0281$
         e $P(Z \le -1.91) = .0281$
         f $P(.35 \le Z \le 1.75) = P(Z \le 1.75) - P(Z \le .35) = .9599 - .6368 = .3$

6.2-3   **a** $P(X < 0) = P[(X - 8)/3.5 < (0 - 8)/3.5] = P(Z < -2.29) = 1 - P(Z \geq -2.29)$
             $= 1 - .9890 = .011$

         b $P(-2 \leq X \leq 2) = P[(-2 - 8)/2.5) \leq Z \leq (2 - 8)/3.5)] = P(-2.866 \leq Z \leq -1.71)$
             $= (1 - .9564) - (1 - .9979) = .0415$

         c $P(X > 15) = P[Z > (15 - 8)/3.5] = P(Z > 2) = 1 - .9772 = .0228$

6.2-7 $n = 75$ $p = .05$
         $\mu \approx np = 75(.05) = 3.75$
         $\sigma \approx \sqrt{np(1 - p)} = \sqrt{75(.05)(.95)} = 1.89$
         $P(X > 5) = P[Z > (5 - 3.75)/1.89] = P(Z > .66) = 1 - .7454 = .2546$

6.2-9 $P(N \geq 500) = P(\ln(N) \geq 6.21) = P(Z \geq (6.21 - 6.3)/1.6) = P(Z \geq .57)$
         $= 1 - .7157 = .2843$

SECTION 6.3

6.3-1   **a** $\Gamma(5) = 4! = 24$
         b $\Gamma(3/2) = 1/2\Gamma(1/2) = 1/2\sqrt{\pi} = .886$

6.3-3   **a** A chi-square random variable is a gamma random variable with
            $\alpha = n/2$ and $\beta = 2$.
            $P(X \leq 8) = P(Y \leq 8/2) = P(Y \leq 4) = .567$ (since $\alpha = 4$)
         b $E(X) = \alpha\beta = 8,\ VAR(X) = \alpha\beta^2 = 16$

6.3-5   **a** $\sigma^2 = 2n = 1$
         b $P(7 - 2\sqrt{14} < X < 7 + 2\sqrt{14}) = P(0 < X < 14.48) = P(0 < Y < 7.24) = .95$
            Chebychev's inequality states that approximately 95% of the probability
            should fall within 2 standard deviations of the mean, which is agreement
            with the above calculation.

6.3-7 $P(\text{Time} > 9) = P(3 + X > 9) = P(X > 6) = 1 - .849 = .151$, where
         $\alpha = 4$ and $\beta = 1$

SECTION 6.4

6.4-1   **a** $f(t) = \dfrac{\beta}{\alpha}\left(\dfrac{t}{\alpha}\right)^{\beta - 1} e^{-(x/\alpha)\beta}$
         b $E(T) = \alpha\Gamma(1 + 1/\beta) = (.05)\Gamma(1 + 4) = (.05)\Gamma(5) = (.05)(4!) = 1.2$
            $VAR(T) = (.05)^2[\Gamma(1 + 8) - [\Gamma(1 + 4)]^2] = (.05)^2(8! - 4!^2)$
            $= (.05)^2 (40,320 - 576) = 99.4$
         c $P(T > 180) = \exp[-(180/.05)^{.25}] = .0004$

6.4-3 $\dfrac{\mu^2}{\mu^2 + \sigma^2} = .917$, and from table 6.4-1, β is approximately equal to 3.7,
         $\Gamma(1 + 1/\beta) \approx .9025$, and $\alpha = 10/.9025 = 11.08$.

$P(a < X < b) = P(X < b) - P(X < a) = 50 \Rightarrow P(X < b) = .75$ and $P(X < a) = .25$

$P(X < b) = 1 - \exp((-b/11.08)^{3.7}) = .75 \Rightarrow \exp((-b/11.08)^{3.7}) = .25$

$\Rightarrow b = (11.08)(-\ln(.25))^{1/3.7} = 12.1$

$P(X < a) = 1 - \exp((-a/11.08)^{3.7}) = .25 \Rightarrow \exp((-a/11.08)^{3.7}) = .75$

$\Rightarrow a = (11.08)(-\ln(.75))^{1/3.7} = 7.9$

Thus, the middle 50% is contained in the range 7.9 to 12.1 inches.

SECTION 6.5

6.5-1 $\displaystyle\int_0^1 x^5(2x)dx = \int_0^1 2x^6 dx = (2/7)x^7 \Big|_0^1 = 2/7$

6.5-3 $\displaystyle M_x(t) = Ee^{Xt} = \int_0^\infty e^{xt}\lambda e^{-\lambda x}dx = \int_0^\infty \lambda e^{-x(\lambda - t)}dx = \frac{\lambda}{(t-\lambda)}e^{-x(\lambda-t)}\Big|_0^\infty = -\frac{\lambda}{t-\lambda} = \frac{\lambda}{\lambda-t}$

$M_x^{(1)}(t) = \dfrac{\lambda}{(\lambda-t)^2}$ and $M_x^{(2)}(t) = \dfrac{-\lambda}{(\lambda-t)^3}$. Evaluating these derivatives at

$t = 0$, we find $\mu = 1/\lambda$ and $\sigma^2 = 1/\lambda^2 \Rightarrow \mu = \sigma$.

6.5-5 **a** $M_x(t) = Ee^{Xt} = \displaystyle\int_0^\infty \frac{1}{\Gamma(\alpha)\beta^\alpha}e^{tx}x^{\alpha-1}e^{-x/\beta}dx = \frac{1}{\Gamma(\alpha)\beta^\alpha}\int_0^\infty x^{\alpha-1}e^{-x\left(\frac{\beta}{1-\beta t}\right)}dx$

Using the fact that, for any positive constants a and b,

$f(x) = \dfrac{1}{\Gamma(a)b^a}x^{\alpha-1}e^{-x/\beta}$ is a pdf,

we have that $\displaystyle\int_0^\infty \frac{1}{\Gamma(a)b^a}x^{\alpha-1}e^{-x/\beta}dx = 1$ and $\displaystyle\int_0^\infty x^{\alpha-1}e^{-x/\beta}dx = \Gamma(\alpha)b^a$.

Thus, $M_x(t) = \dfrac{1}{\Gamma(\alpha)\beta^\alpha}\Gamma(\alpha)\left(\dfrac{1}{1-\beta t}\right)^\alpha = \left(\dfrac{1}{1-\beta t}\right)^\alpha$.

b $M_1(t) = \dfrac{\alpha\beta}{(1-\beta t)^{\alpha+1}}$ $M_2(t) = \dfrac{\alpha(\alpha-1)\beta^2}{(1-\beta t)^{\alpha+2}}$. Evaluating these

derivatives at $t = 0$, $\mu = \alpha\beta$ and $\sigma^2 = \alpha(\alpha-1)\beta^2 - \alpha^2\beta^2 = \alpha\beta^2$.

SECTION 6.6

6.6-1 The first two moments of a normal random variable are found by setting

$M_1 = E(X) = \mu = \overline{X}$ and $M_2 = E(X^2) = \sigma^2 + \mu^2$

Thus, $\hat\mu = \overline{X}$ and $\hat\sigma^2 = M_2 - \mu^2 = (\Sigma X_i^2/n) - \overline{X}^2 = [\Sigma X_i^2 - n\overline{X}]/n = S^2$

6.6-3 **a** Since X is exponential, $E(X) = 1/\lambda \Rightarrow \hat\lambda = 1/\overline{X}$

b $\hat\lambda = .4503$; $\hat\sigma^2 = 1/\hat\lambda^2 = 4.930$

6.6-5 $M_1 = \overline{X} = 1.1455$

$M_2 = \Sigma X_i^2/n = 1.6578$

$\hat{\alpha} = \dfrac{(M_1)^2}{M_2 - (M_1)^2} = 3.7964$

$\hat{\beta} = (M_1 / \hat{\alpha}) = .3017$

CHAPTER 7
COUNTING AND QUEUING PROCESSES

SECTION 7.1

7.1-3 $P = \lambda\Delta \Rightarrow \Delta = P/\lambda \Rightarrow \Delta = (.05 \text{ customer/frame})/(20 \text{ customers/hour})$
$= .0025 \text{ hour/frame}$
Thus, the frame length is .0025 hour/frame.
The number of frames per hour is $1/\Delta = 400$ frames/hour.

7.1-5   **a**   Let Y be the number of 30-second frames from one defect to the next.
Then Y has a geometric distribution. By Theorem 7.1-2,
$P(Y = y) = (1 - p)^{y-1} p = (.98)^{y-1} (.02), \; y = 1, 2,...$
From example 7.1-5 we know that $P(Y > y) = (1 - p)^y = .98^y$
So $P(Y > 40) = .98^{40} = .44570$
   b $\Delta = 1/120$
$p = \lambda\Delta \Rightarrow \lambda = p/\Delta = (.02)/(1/120) = (120)(.02) = 2.4$
   c   Let T be the time from 1 defect to the next. By theorem 7.1-3,
$E(T) = 1/\lambda = 1/2.4 = .4167$ hour, or about 25 minutes
   d $E(T) = 1/24 = .4167$ hour, or 25 minutes
$VAR(T) = (1 - .02)/(.02)^2 = 2450 \Rightarrow STD(T) = 49.5$ frames
or 24.75 minutes
Alternately, $VAR(T) = [1 - 2.4(1/120)]/(2.4)^2 = .1702$
$\Rightarrow STD(T) = .4125$ hour $= 24.75$ minutes
A 95% confidence interval is given by $25 \pm 2(24.75) \approx (0, 75.5)$

SECTION 7.2

7.2-1   **a** $\lambda = 5/30$ days; $t = 15$ days
$\mu = \lambda t = (5/30 \text{ days})(15 \text{ days}) = 2.5$
$P(N(15) > 5) = 1 - P(N(15) \leq 5) = 1 - \sum\limits_{x=0}^{5} \dfrac{e^{-2.5}(2.5)^x}{x!} = 1 - .95797 = .04203$
   b $t = 7$, so $\mu = \lambda t = (5/30 \text{ days})(7 \text{ days}) = 1.16667$
$\Rightarrow P(N(7) = 0) = \dfrac{e^{-.16667}(.16667)^0}{0!} = .31140$

7.2-3   **a** $P(N(1) > 2) = 1 - P(N(1) \leq 2) = 1 - \sum\limits_{x=0}^{2} \dfrac{e^{-.5}(.5)^x}{x!} = 1 - .98652 = .01438$
   b $P(T > 4) = P(N(4) = 0) = \dfrac{e^{-(.05)(4)}[(.5)(4)]^0}{0!} = .13534$

7.2-5 λ = 4 customers/hour
Expected number of customers in 8 hours is λ(8 hours)
= (4 customers/hour)(8 hours) = 32
The standard deviation of this Poisson process is $\sqrt{32}$ = 5.66,
so a likely range for the number of customers visiting the booth in 8 hours
is 32 $\pm$ 2(5.66) = (20.67, 33.31).

7.2-7 λ = 9 accesses/week
$P(N(2) \geq 30)$ = .0059, so it appears that 30 accesses by one user is a bit out of
the ordinary.

SECTION 7.3

7.3-3 $f_s(t) = \dfrac{t^4 e^{-t}}{4!}$, μ = 5, σ = $\sqrt{5}$

7.3-5 λ = .2/hr, t = 6 hr
$\mu = \lambda t$ = (.2 surges/hour)(6 hours) = 1.2 surges
$$P(W_s > 6) = P(N(6) \leq 3) = \sum_{x=0}^{3} \frac{e^{-1.2}(1.2)^x}{x!} = .9662$$

SECTION 7.4

7.4-1   **a** $\lambda_s = 1/\mu_s$ = 32/5 services/day
b $P_A = \lambda_A \Delta$ = (4 arrivals/day)(1/32 days/frame) = 1/8 arrivals/frame
= .0125 arrivals/frame
$P_s = \lambda_s \Delta$ = (32/5 services/day)(1/32 days/frame) = 1/5 services/frame
= .2 services/frame

c   Day 1

Frame	U_s	U_A	IS	IA	Q
0	*	*	0	0	0
1	*	.048	0	1	1
2	.681	.391	0	0	1
3	.257	.641	0	0	1
4	.879	.627	0	0	1
5	.959	.299	0	0	1
6	.736	.280	0	0	1
7	.910	.189	0	0	1
8	.948	.356	0	0	1
9	.334	.887	0	0	1
10	.395	.070	0	1	2
11	.410	.840	0	0	2
12	.333	.317	0	0	2

13	.935	.205	0	0	2
14	.020	.854	1	0	1
15	.973	.617	0	0	1
16	.167	.427	1	0	0
17	.700	.080	0	1	1
18	.103	.540	1	0	0
19	.333	.034	0	1	1
20	.927	.857	1	0	1
21	.082	.513	0	0	0
22	.602	.950	0	0	0
23	.586	.010	0	1	1
24	.143	.741	1	0	0
25	.242	.873	0	0	0
26	.074	.964	0	0	0
27	.264	.664	0	0	0
28	.264	.943	0	0	0
29	.773	.562	0	0	0
30	.553	.886	0	0	0
31	.129	.301	0	0	0
32	.491	.496	0	0	0

b Day 2

Frame	U_s	U_A	IS	IA	Q
0	*	*	0	0	0
1	*	.048	0	1	1
2	.681	.391	0	0	1
3	.257	.641	0	0	1
4	.879	.627	0	0	1
5	.959	.299	0	0	1
6	.736	.280	0	0	1
7	.910	.189	0	0	1
8	.948	.356	0	0	1
9	.334	.887	0	0	1
10	.395	.070	0	1	2
11	.410	.840	0	0	2
12	.333	.317	0	0	2
13	.935	.205	0	0	2
14	.020	.854	1	0	1
15	.973	.617	0	0	1
16	.167	.427	1	0	0
17	.700	.080	0	1	1
18	.103	.540	1	0	0
19	.333	.034	0	1	1

20	.927	.857	1	0	1
21	.082	.513	0	0	0
22	.602	.950	0	0	0
23	.586	.010	0	1	1
24	.143	.741	1	0	0
25	.242	.873	0	0	0
26	.074	.964	0	0	0
27	.264	.664	0	0	0
28	.264	.943	0	0	0
29	.773	.562	0	0	0
30	.553	.886	0	0	0
31	.129	.301	0	0	0
32	.491	.496	0	0	0

SECTION 7.5

7.5-1

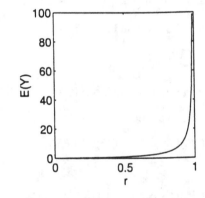
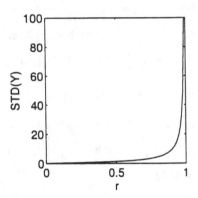

7.5-3 $P(X \le 2) = .90 \Rightarrow 1 - r^3 = .90 \Rightarrow r^3 = .10 \Rightarrow .4642$
$\lambda_A / \lambda_S = .4642 \Rightarrow \lambda_S = 21.5443$

7.5-7 **a** $\mu = r/(1 - r) = (10/11)/(1/11) = 10$
b $\mu = 5 \Rightarrow r/(1 - r) = 5 \Rightarrow r = 5 - 5r \Rightarrow 6r = 5 \Rightarrow r = 5/6$
Thus, $\lambda_A/\lambda_S = 5/6 \Rightarrow \lambda_S = (6/5)(10 \text{ cars/day}) = 12 \text{ cars/day}$

7.5-9 $\lambda_A = 8$ cases/hr, $\lambda_S = 10$ cases/hr, $r = .8$
$P(X \ge 3) = P(X > 2) = r^3 = .8^3 = .512$

SECTION 7.6

7.6-1 $P(0 \to 0) = P(0 \text{ arrivals}) = 1 - P_A = .90$
$P(0 \to 1) = P(1 \text{ arrival}) = P_A = .10$
$P(1 \to 0) = P(0 \text{ arrivals and 1 service}) = (1 - P_A)P_S = (.90)(.15) = .135$

33

$P(1 \to 1) = P(1 \text{ arrival and } 1 \text{ service}) + P(0 \text{ arrivals and } 0 \text{ services})$
$= P_A P_S + (1 - P_A P_S) = (.1)(.15) + (.90)(.85) = .78$

$P(1 \to 2) = P(1 \text{ arrival and } 0 \text{ services}) = P_A(1 - P_S) = (.10)(.85) = .085$

$P(2 \to 0) = P(0 \text{ arrivals and } 2 \text{ services}) = (1 - P_A)P_S^2 = (.90)(.15)^2 = .0202$

$P(2 \to 1) = P(0 \text{ arrivals and } 1 \text{ service}) + P(1 \text{ arrival and } 2 \text{ services})$
$= (1 - P_A)\begin{pmatrix} 2 \\ 1 \end{pmatrix} P_S(1 - P_S) = P_A P_S^2 = (.9)(2)(.15)(.85) + (.10)(.15)^2 = .2317$

$P(2 \to 2) = P(0 \text{ arrivals and } 0 \text{ services}) + P(1 \text{ arrival and } 1 \text{ service})$
$= (1 - P_A)(1 - P_S)^2 + P_A\begin{pmatrix} 2 \\ 1 \end{pmatrix} P_S(1 - P_S) = (.9)(.85)^2 + (.10)(2)(.15)(.85)$
$= .6757$

$P(2 \to 3) = P(1 \text{ arrival and } 0 \text{ services}) = P_A(1 - P_S)^2 = (.10)(.85)^2 = .0722$

For $i \geq 3$,
$P(i \to i - 3) = P(0 \text{ arrivals and } 3 \text{ services}) = (1 - P_A)P_S^3 = (.90)(.15)^3 = .0030$

$P(i \to i - 2) = P(0 \text{ arrivals and } 2 \text{ services}) + P(1 \text{ arrival and } 3 \text{ services})$
$= (1 - P_A)\begin{pmatrix} 3 \\ 2 \end{pmatrix} (.15)^2(.85) + (.10)(.15)^3 = .0519$

$P(i \to i - 1) = P(0 \text{ arrivals and } 2 \text{ services}) + P(1 \text{ arrival and } 2 \text{ services}) = .2983$

$P(i \to i) = P(0 \text{ arrivals and } 0 \text{ services}) + P(1 \text{ arrival and } 1 \text{ service})$
$= (1 - P_A)(1 - P_S)^3 + P_A\begin{pmatrix} 3 \\ 1 \end{pmatrix} P_S(1 - P_S)^2 = (.90)(.85)^3 + (.10)(3)(.15)(.85)^2$
$= .5852$

$P(i \to i + 1) = P(1 \text{ arrival and } 0 \text{ services}) = P_A(1 - P_S)^3 = (.10)(.85)^3 = .0614$

The transition matrix is given by

$$
P = \begin{array}{c} 0 \\ 1 \\ 2 \\ 3 \\ 4 \\ 5 \\ \vdots \end{array}
\begin{bmatrix}
.90 & .10 & 0 & 0 & 0 & 0 & \cdots \\
.135 & .780 & .085 & 0 & 0 & 0 & \cdots \\
.0202 & .2317 & .6757 & .0722 & 0 & 0 & \cdots \\
.0030 & .0519 & .2983 & .5852 & .0614 & 0 & \cdots \\
0 & .0030 & .0519 & .2983 & .5852 & .0614 & \cdots \\
0 & 0 & .0030 & .0519 & .2983 & .5852 & \cdots \\
\vdots & \vdots & \vdots & \vdots & \vdots & \vdots & \ddots
\end{bmatrix}
$$

SECTION 7.7

7.7-1 Using the formulas $\pi_0 = \left[1 + \sum_{j=1}^{k-1} r^j/j! + \sum_{j=k}^{\infty} (r^j)/ [(k!k^{j-k}] \right]$

$\pi_j = \pi_0 (r^j/j!), j < k$ and $\pi_j = \pi_0 (r/k)^{j-k}, j \geq k$, we obtain

r = .5			r = .9	
j	π_j		j	π_j
0	.6065		0	.406211
1	.3032		1	.365590
2	.0758		2	.164515
3	.0126		3	.049355
4	.0016		4	.011105
5	.0002		5	.002499
6	.00002		6	.000562
7	0		7	.000126
8	0		8	.000028
9	0		9	.000006
10	0		10	.000001

Thus, for r = .9, E(N) = .90 and STD(N) = .96 and for r = .5, E(N) = .5 and STD(N) = .71.

7.7-7 For this system, λ_A = 8 calls/hour, λ_S = 3 services/hour, and r = 2.6667. Thus, a table of probabilities is given by

j	π_j
0	.0671
1	.1789
2	.2386
3	.2121
4	.1414

Thus, P(a counselor is free) = P(4 or fewer callers in system)
= .0671 + .1789 + .2386 + .2121 + .1414 = .8381

SECTION 7.8

7.8-1 **a** $a_0 = a_1 = a_2 = a_3$ = .2 cars/day
a_k = 0 cars/day for k $\geq$ 4
$s_1 = s_2 = s_3 = s_4$ = .25 cars/day

$$\pi_0 = .297$$
$$\pi_1 = .238$$
$$\pi_2 = .190$$
$$\pi_3 = .152$$
$$\pi_4 = .122$$

b $\mu = \sum_{j=0}^{4} j\pi_j = 1.562, \ \sigma^2 = \sum_{j=0}^{4} (j - \mu)\pi_j = 1.877, \ \sigma = \sqrt{1.877} = 1.370$

7.8-3 $\lambda_A = 8$ cust/min, $\lambda_S = 3$ cust/min, and $r = 8/3$.

$\pi_0 = e^{-8/3}$, $\pi_j = e^{-8/3}(8/3)^j/j!$ for $j = 1, 2, 3,...$

The number in the system has a Poisson distribution with mean $\mu = r = 8/3$.

CHAPTER 8
THE DISTRIBUTION OF
SUMS OF RANDOM VARIABLES

SECTION 8.1

8.1-1 **a** Assuming the runners' times are independent of one another, we get
$E(T) = E(X_1 + X_2 + X_3 + X_4) = E(X_1) + E(X_2) + E(X_3) + E(X_4)$
$= 54 + 56 + 52 + 51 = 213$
$VAR(T) = VAR(\sum_{i=1}^{4} X_i) = \sum_{i=1}^{4} VAR(X_i) = 1.5^2 + 2.0^2 + 1.5^2 + 1.0^2$
$= 9.5$ seconds
$STD(T) = \sqrt{VAR(T)} = 3.08$ seconds

b $E(T) \pm 2STD(T) = 213 \pm 2(3.08) = [206.84, 216.16]$

c $P(T < 204) = P(Z < (204 - 213)/3.08] = P(Z < -2.92) = .5000 - .4982 = .0018$

8.1-3 Let T be the total time to rig a sail. Then T is normal with a mean of 10.50 and a standard deviation of 2.958. $P(T > 15) = P(Z > (15 - 10.5)/2.958)$
$= P(Z > 1.52) = .5000 - .4357 = .0643$

8.1-5 $M_X(t) = (1 - t/\lambda)^{-1}$
Let $Y = \sum_{i=1}^{n} X_i$. Then by theorem 8.1.2, $M_Y(t) = \prod_{i=1}^{n} M_{X_i}(t) = (1 - (1/\lambda)t)^{-n}$, $t < \lambda$
which is the moment generating function of a gamma(n, $1/\lambda$) random variable.

SECTION 8.2

8.2-1 **a** Let X_i, i = 1,...,20 be the lifetime of the ith bulb. $X_i \sim$ exponential(900).
Then $T = \sum_{i=1}^{n} X_i$.
Assuming that the lifetimes are independent, we have
$E(T) = E(\sum_{i=1}^{20} X_i) = \sum_{i=1}^{20} 900 = 18,000$
$VAR(T) = VAR(\sum_{i=1}^{20} X_i) = \sum_{i=1}^{20} VAR(X_i) = \sum_{i=1}^{20} 900^2 = (20)(900)^2$
$\Rightarrow STD(T) = 900 \sqrt{20} = 4024.92$

b By the central limit theorem, $T \sim N(18,000, 4024.92)$.
So $P(T > 22,000) = P(Z > (22,000 - 18,000)/4024.92) = P(Z > .99)$
$= .5000 - .3389 = .1611$

c $P(T < 16,000) = P(Z < (16,000 - 18,000)/4024.92) = P(Z < -.50)$
$= .5000 - .1915 = .3085$

8.2-3 **a** Let X_i be the profit on day i. By the central limit theorem,

$$T = \sum_{i=1}^{30} X_i \sim N(9000, 273.86)$$

Thus, $P(T > 9500) = P(Z > (9500 - 9000)/273.86) = P(Z > 1.83)$
$= .5000 - .4664 = .0336$

b By the central limit theorem, $\overline{X} \sim N(300, 50/\sqrt{30}) = N(300, 9.13)$
$P(\overline{X} < 290) = P(Z < (290 - 300)/9.13) = P(Z < -1.10)$
$= .5000 - .3643 = .1357$

8.2-9 By the central limit theorem, $\hat{p} \sim N(.30, \sqrt{(.30)(.70)/400}) = N(.30, .0229)$
$P(\hat{p} < .28) = P(Z < (.28 - .30)/.0229) = P(Z < -.87) = .5000 - .3078 = .1922$

SECTION 8.3

8.3-1 $\overline{X} \pm Z_{.025} \sigma_x = .9 \pm 1.96(.2/\sqrt{35}) = .9 \pm .0663 = [.834, .966]$

8.3-5 $\hat{p} \pm z_{.025} \hat{\sigma}_p = .22 \pm 1.96\sqrt{(.22)(.78)/150} = .22 \pm .0663 = [.1537, .2863]$

SECTION 8.4

8.4-1 $\lambda = \$9/hr, \mu = \$40, \sigma = \$12, t = 8$
$E(X(8)) = (\$9/hr)(8\ hr)(\$40) = \$2880$
$VAR(X(8)) = (\$9/hr)(8\ hr)(\$40)^2 + (\$9/hr)(8\ hr)(\$12)^2 = 125,568$
$STD(X(8)) = \sqrt{125,568} = 354.36$

8.4-3 **a** $\lambda = 15/hr, \mu = 1.5\ hr, \sigma = .25$
$E(X(8)) = (15)(8)(1.5) = 180$
$VAR(X(8)) = (15)(8)(1.5)^2 + (15)(8)(.25)^2 = 277.5$
$STD(X(8)) = 16.65$

b $P(X(8) < 200) = P(Z < (200 - 180)/16.65) = P(Z < 1.20) = .8849$

CHAPTER 9
SELECTED SYSTEMS MODELS

SECTION 9.1

9.1-1 **a** Let T_i equal the time it takes the ith hiker to get to the destination.

$F_{T_i}(t) = (t - 35)/25, \quad 35 < t < 60$

$T = \max(T_1,...,T_5)$

$F_{max}(t) = [(t - 35)/25]^5, \quad 35 < t < 60$

b $f_{max}(t) = (5)[(t - 35)/25]^4 (1/25) = (1/5)[(t - 35)/25]$

$E(T_{max}) = \int_{35}^{60} (t/5)[(t - 35)/25]^4 dt$

Let $w = (t - 35)/25, \quad t = 25w + 35,$ and $dt = 25dw$

$E(T_{max}) = \int_0^1 [(25w + 35)/5](w)^4 (25)dw = \int_0^1 (125w^5 + 175w^4)dw$

$= (125/6)w^6 + (175/5)w^5 \; |_0^1 = (125/6) + (175/5) = 55.8$

9.1-3 $f(t) = (4/5)e^{-t/5} (1 - e^{-t/5})^3 \;\Rightarrow\; E(T) = \int_0^\infty t(4/5)e^{-t/5} (1 - e^{-t/5})^3 dt$

Integration by parts: Let $u = t, \quad du = dt, \quad v = -(1 - e^{-t/5})^4,$

$dv = (4/5)e^{-t/5} (1 - e^{-t/5})^3 \, dt$

Then $E(T) = -t(1 - e^{-t/5})^4 \; |_0^\infty + \int_0^\infty (1 - e^{-t/5})^4 dt = \int_0^\infty (1 - e^{-t/5})^4 dt$

Note that $\int_0^\infty (1 - e^{-t/5})^n \, dt = \int_0^\infty (1 - e^{-t/5})^{n-1} (1 - e^{-t/5}) dt$

$= \int_0^\infty (1 - e^{-t/5})^{n-1} dt + \int_0^\infty (1 - e^{-t/5})^{n-1} e^{-t/5} dt$

where $\int_0^\infty (1 - e^{-t/5})^{n-1} e^{-t/5} dt = 5\int_0^1 u^{n-1} \, du = 5(u^n /n) \, |_0^1 = 5/n$

Continuing in a similar way, we obtain the recursive formula

$\int_0^\infty (1 - e^{-t/5})^n \, dt = \sum_{i=1}^n (5/i).$

Thus $E(T) = \int_0^\infty (1 - e^{-t/5})^4 \, dt = 5/4 + 5/3 + 5/2 + 5/1 = 10.4$

SECTION 9.3

9.3-1

$$P = \begin{array}{c} 1 \\ 2 \\ 3 \\ 4 \end{array} \begin{bmatrix} -2 & 2 & 0 & 0 \\ 1 & -3 & 2 & 0 \\ 0 & 1 & -3 & 2 \\ 0 & 0 & 1 & -1 \end{bmatrix}$$

Solve the system of equations:

$-2\pi_1 + \pi_2 = 0$
$2\pi_1 - 3\pi_2 + \pi_3 = 0$
$2\pi_2 - 3\pi_3 + \pi_4 = 0$
$2\pi_3 - \pi_4 = 0$
$\pi_1 + \pi_2 + \pi_3 + \pi_4 = 1$

$\pi_2 = 2\pi_1$
$2\pi_1 - 6\pi_1 = -\pi_3$ or $\pi_3 = 4\pi_1$
$\pi_4 = 2\pi_3 = 8\pi_1$

$\pi_1 + 2\pi_1 + 4\pi_1 + 8\pi_1 = 1 \Rightarrow \pi_1 = 1/15,\ \pi_2 = 2/15,\ \pi_3 = 4/15,\ \pi_4 = 8/15$

9.3-3

$$P = \begin{array}{c} 0 \\ 1 \\ 2 \\ \vdots \\ 9 \\ 10 \end{array} \begin{bmatrix} -5 & 5 & 0 & \cdots & 0 & 0 \\ 0 & -5 & 5 & \cdots & 0 & 0 \\ 0 & 0 & -5 & \cdots & 0 & 0 \\ \vdots & \vdots & \vdots & \ddots & \vdots & \vdots \\ 0 & 0 & 0 & \cdots & -5 & 5 \\ 2 & 0 & 0 & \vdots & 0 & -2 \end{bmatrix}$$

$-5\pi_0 + 2\pi_{10} = 0 \qquad \Rightarrow \qquad \pi_{10} = (5/2)\pi_0$
$5\pi_0 - 5\pi_1 = 0 \qquad \Rightarrow \qquad \pi_0 = \pi_1$
$5\pi_1 - 5\pi_2 = 0 \qquad \Rightarrow \qquad \pi_1 = \pi_2$

$5\pi_8 - 5\pi_9 = 0 \qquad \Rightarrow \qquad \pi_8 = \pi_9$
$5\pi_9 - 2\pi_{10} = 0 \qquad \Rightarrow \qquad \pi_{10} = (5/2)\pi_9$

$\pi_0 + \pi_1 + \ldots + \pi_9 + \pi_{10} = 1$
$\pi_0 + \pi_0 + \ldots + \pi_0 + (5/2)\pi_0 = 1$
$(9 + 5/2)\pi_0 = 1 \Rightarrow \pi_0 = 2/23$
$\pi_0 = \pi_1 = \ldots = \pi_9 = 2/23$
$\pi_{10} = 5/23$

9.3-7   Description: This program simulates the movement of a robot arm. The amount of time that the robot spends at station 1 is exponential with mean 1/8, the amount of time that the robot spends at station 2 is exponential with mean 1/9, and the amount of time that the robot spends at station 3 is exponential with mean 1/10. The conditional transition probabilities were calculated in example 9.3-3.

Variables:

N   The transition number

$T(\cdot)$ The time at which the Nth transition occurs

$S(\cdot)$ The state to which the process goes when the Nth transition
               is made

Initialize:

N = 1, S(0) = 1, T = 0

Algorithm:

Repeat 1000 times

1)   Generate two random numbers, u and v

2)   If S(N - 1) = 1 then set T(N) = T(N - 1) - 1/8[ln(1 - u)]
         else if S(N - 1) = 2 then set T(N) = T(N - 1) - 1/9[ln(1 - u)]
         else if S(N - 1) = 3 then set T(N) = T(N - 1) - 1/10[ln(1 - u)]

3)   If S(N - 1) = 1 and v < 5/8 then set S(N) = 2
         else if S(N - 1) = 1 and v > 5/8 then set S(N) = 3
         else if S(N - 1) = 2 and v > 5/9 then set S(N) = 1
         else if S(N - 1) = 2 and v > 5/9 then set S(N) = 3
         else if S(N - 1) = 3 and v < 4/10 then set S(N) = 1
         else if S(N - 1) = 3 and v > 4/10 then set S(N) = 2

4)   Increment N

41

CHAPTER 10
THE RELIABILITY FUNCTION

SECTION 10.1

10.1-1 $F(t) = (1/5)t, \ 0 \le t \le 5$
$R(t) = 1 - (1/5)t, \ 0 \le t \le 5$

10.1-3 $R(t) = P(\text{both survive}) = R_1(t)R_2(t) = [\exp(t/8)^2]^2 = \exp[(-2(t/8)^2)] = \exp(-t^2/32)$
Thus, $R(t) = \exp(-t^2/32), \ 0 \le t < \infty$

10.1-5 $R(t) = [1 - (1 - e^{-3.5t}R_{34}(t))(1 - e^{-4.1t}R_{56}(t))]e^{-3.8t}$
where $R_{34}(t) = 1 - (1 - e^{-2.1t})(1 - e^{-2.4t})$ and $R_{56}(t) = 1 - (1 - e^{-1.9t})(1 - e^{-3.1t})$

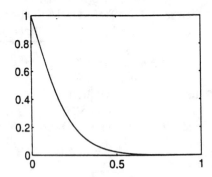

10.1-7 $P(\text{at least one survives}) = R_{max}(t) = 1 - (1 - e^{-t/2000})^2 = .995$
$\Rightarrow 1 - e^{-t/2000} = \sqrt{.005} \Rightarrow e^{-t/2000} = .92929 \Rightarrow t = -2000\log(.92929)$
$= 146.7$ hours

SECTION 10.2

10.2-1 $h(t) = f(t)/R(t)$, where $R(t) = 1 - (1 - e^{-t})(1 - e^{-2t})(1 - e^{-3t})$ and
$f(t) = e^{-t}(1 - e^{-2t})(1 - e^{-3t}) + 2e^{-2t}(1 - e^{-t})(1 - e^{-3t}) + 3e^{-3t}(1 - e^{-t})(1 - e^{-2t})$

10.2-3 $F(t) = (1/4)(t - 1), \ 1 \le t \le 5$
$R(t) = 1 - (1/4)(t - 1), \ 1 \le t \le 5$
$f(t) = 1/4, \ 1 \le t \le 5$
$h(t) = f(t)/R(t) = (1/4)/(1 - (1/4)(t - 1)) = 1/(4 - (t - 1)) = 1/(5 - t)$

10.2-5 $R(t) = \exp(-\int_0^t h(s)ds) = \exp(-\int_0^t 2e^{-.5s}ds) = \exp(-4(e^{-t/2} - 1))$

$R(1) = \exp(-4(e^{-.5} - 1)) = .075$

SECTION 10.3

10.3-1   **a**   P(12 or more requests) = P(3 or more messages) = $\sum_{k=12}^{\infty} e^{-10}10^k/k! = 1 - .7916$

     b   P(time between dispatches $\leq 1/2$) = P(4 or more dispatches in 1/2 hour)

        $= \sum_{k=4}^{\infty} e^{-5}5^k/k! = 1 - .2650 = .7350$

SECTION 10.4

10.4-1 $\mu_u = (8)(2) = 16$, $\mu_D = (.25)(2) = .5$

$P(\text{line is up}) = 16/(16 + .5) = .97$

10-4-3 $\pi_0 = 1/[1 + \lambda/\mu + (1/2)(\lambda/\mu)] = 1/[1 + 10 + 50] = 1/61$

$\pi_1 = 10/61$

$\pi_2 = 51/61$

SECTION 10.5

10.5-1   **a**   P(at least one mistake) = 1 - P(no mistakes) = $\exp(-(.5)^2) = .4187$

     b $m(2) = 2^2 = 1.15$

        $P(N(2) \leq 3) = \sum_{k=0}^{3} e^{-1.15}(1.15)^k/k! = .97$

     c   P(at least one mistake) = 1 - P(no mistakes) = $1 - \exp[-(m(4) - m(3))]$

        $= 1 - \exp(-4^2 + 3^2) = .07$

     d $1 - \exp[-(t + 1)^2 + t^2] = .05 \Rightarrow \exp[-(t + 1)^2 + t^2] = .95$

        $t^2 - (t + 1)^2 = \log(.95) = -.05129$

        $(t + 1)^2 - t^2 = .05129$

        $t = 5$ days